•DISCOVERING•
JUPITER

The Amazing
Collision in Space

by Melvin Berger

illustrated by Tom Leonard

Scholastic Inc.
New York   Toronto   London   Auckland   Sydney

ISBN: 0-590-48824-4

Text copyright © 1995 by The Melvin H. and Gilda Berger Trust.
Illustrations copyright © 1995 by Scholastic Inc.
All rights reserved. Published by Scholastic Inc.

12 11 10 9 8 7 6 5 4 3 2 1   5 6 7 8 9 0/09

Printed in the U.S.A.   23
First Scholastic Printing, November 1995

Book design by Laurie Williams

·DISCOVERING·
JUPITER

Introduction

Jupiter is a planet, just as Earth is a planet. A planet is a large body in space. There are nine planets in our Solar System. All the planets of the Solar System move in paths, or *orbits*, around the Sun.

The Sun is at the center of the Solar System. The Sun is a star. Stars are much bigger than planets. And stars produce tremendous amounts of heat and light. Planets produce no light and very little heat. The planets of the Solar System get all their light and almost all their heat from the Sun.

As the earth travels in space, one side is lit by the Sun and one side is dark.

Earth, Jupiter, and all the other planets get their light from the Sun. This photograph of the Sun was taken in space. It shows a huge solar flare bursting across the surface.

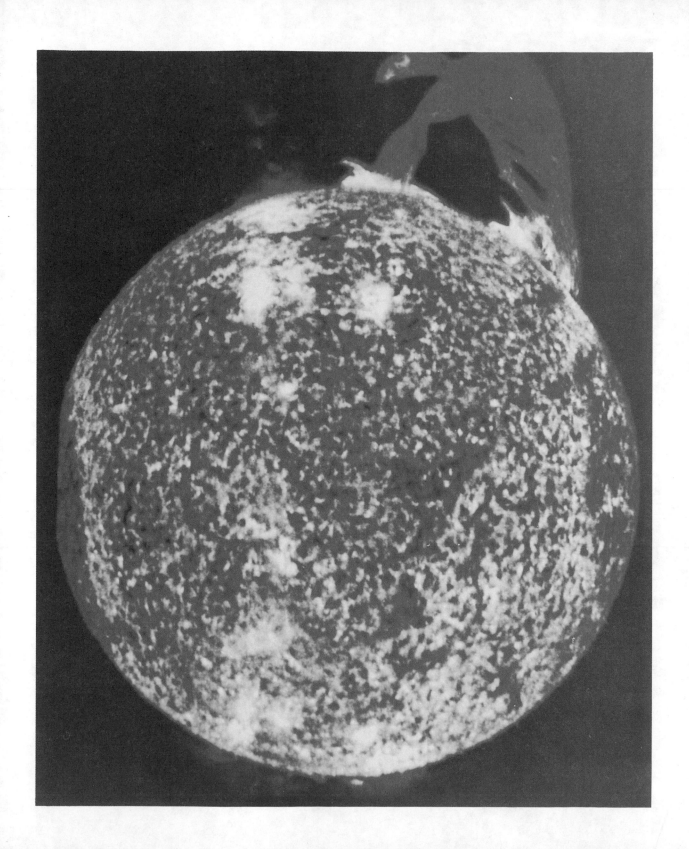

PLUTO

Besides the Sun and planets, the Solar System also includes:

- moons — smaller bodies in orbits around the planets,

- asteroids — planetlike objects in orbits around the sun,

- meteoroids — chunks of stone or metal flying through space,

- comets — balls of ice, metal, and rock with long tails of gas and dust.

SATURN

ASTEROID BELT

URANUS

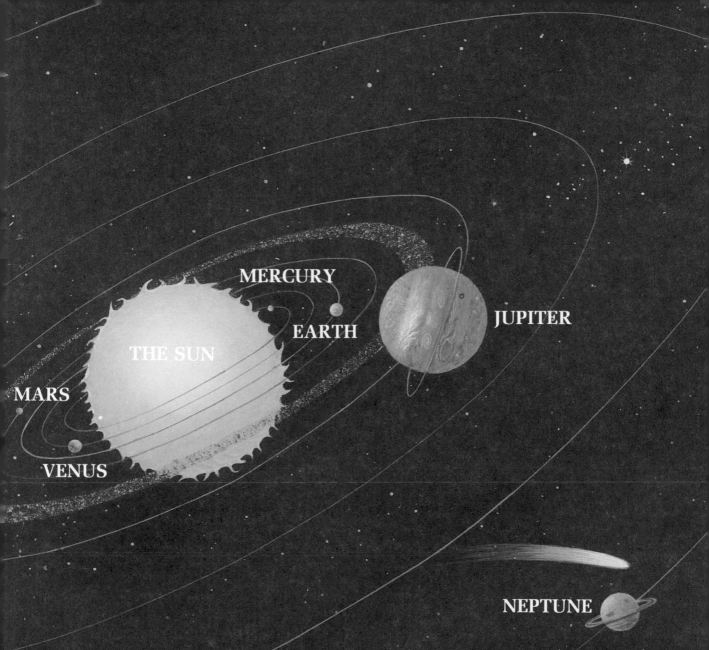

MARS

VENUS

THE SUN

MERCURY

EARTH

JUPITER

NEPTUNE

Nine planets and the Sun make up our Solar System —
Sun, Mercury, Venus, Earth, Mars, Jupiter, Saturn,
Uranus, Neptune, and Pluto.

In the summer of 1994, something amazing occurred in the Solar System. A giant comet slammed into the planet Jupiter! During the next six days, 20 more pieces of the comet struck. Each one caused a huge fireball to explode over Jupiter.

Comets have been smashing into planets for over three billion years. Why was this crash amazing?

It was special because for the first time scientists knew about the collision *before* it happened. People on Earth could watch the explosions through their telescopes. And pictures of the colliding comet could be taken by satellites in space.

And so, in the summer of 1994, all eyes turned toward Jupiter, a planet that has been studied and wondered about for hundreds of years.

Before reading more about this astounding accident in space, here are some things to know about Jupiter.

This is the first true-color photograph of Jupiter, taken in 1991. Jupiter's colors and features change from year to year, so pictures taken now would look a little different.

How Jupiter Got Its Name

Jupiter is a giant planet. It is the largest planet of the Solar System. Even the earliest stargazers could tell the planet was immense. They gave it the name Jupiter, after the all-powerful king of the gods in ancient Roman myths.

Jupiter was believed to be the supreme ruler of the universe. From high in the sky he watched what was happening on Earth. If he saw people being cruel, jealous, or greedy he punished them. His main weapon was the thunderbolt. When necessary, he could also send fierce storms and brutal winds to Earth.

A JUPITER MYTH

The ancient people said that Jupiter would sometimes come down to Earth. He would arrive disguised as a human being.

On one trip, Jupiter dressed himself in the clothes of a poor wanderer. He knocked at many houses asking for food and shelter. Everybody barred his way, except for one extremely poor man and his wife. Although the couple had very little to spare, they fed and sheltered the shabby stranger.

Jupiter thanked the lowly pair and went back to his home in the sky. But he was angry at the cruelty of the other people on Earth. He decided to punish them for their selfishness. He ordered torrents of rain that completely flooded the Earth.

After a long while, the deluge ended. Gone were all the houses and all the coldhearted people that had lived in them. Only the old woman and man who had opened their home to Jupiter were left on Earth.

But the two old people were very lonely. They prayed to the gods for help. A voice told them to walk together and throw stones ahead of them. As they did, the stones changed into human beings. In this way, people returned to the Earth.

Jupiter: Giant of the Solar System

In time, people stopped believing in the forces of the mighty god Jupiter. But they kept the name for this mammoth planet.

How big is Jupiter? Just think:

- A tunnel drilled from the North Pole to the South Pole through the center of Earth would be about 8,000 miles long. But a tunnel drilled through Jupiter would be **90,000** miles long — over ten times as long!

● A plane trip around the Earth's Equator, which circles the planet halfway between the poles, takes about 40 hours. On Jupiter the same journey would take 235 hours!

● Suppose you were the god Jupiter and could hollow out your planet. Within this enormous planet, you could fit 1,000 planets the size of Earth!

Because Jupiter is so much larger than Earth, the pull of gravity is much stronger. An astronaut on Jupiter would weigh more than 2½ times as much as on Earth. Let's say you weigh 80 pounds on Earth. You would weigh over 200 pounds on Jupiter!

Object of Wonder

From the beginning of human history, people looked up in wonder at the sky. They made up stories about the Sun and the stars. In time, some began to study the objects they saw in the heavens. Scientists who study the skies are called astronomers.

At first, astronomers observed the stars and planets with their eyes. Around the year 1600, however, astronomers gained a new tool. It was the telescope. Telescopes collect the dim light from the stars and planets. With the telescope, astronomers could see objects in space much better than before.

The Italian astronomer Galileo Galilei discovered Jupiter with this telescope almost 400 years ago.

The stars are much farther away than planets. They look just like points of light — even through a telescope. But the planets look much clearer and brighter through a telescope. They also look bigger. This helps astronomers see more details on the surfaces of the planets.

Today, most astronomers don't only observe objects in the Solar System through telescopes. They hook up cameras to the telescopes and take pictures. Then they study the photos to make new observations.

Telescopes today are very powerful. One of the largest telescopes ever built is the Hale Telescope in Pasadena, California.

The Hale Telescope is so big that a worker needs a special lift to reach the lenses.

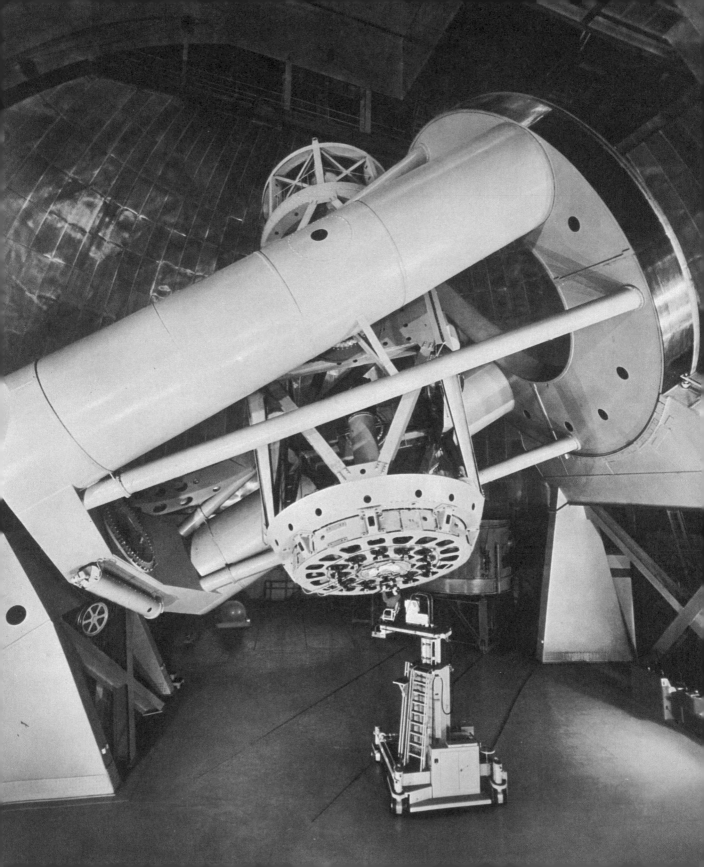

Telescopes, with and without cameras, have been the main tool of astronomers for nearly 400 years. But for the last 40 years astronomers have had other helpful tools to use. They are called space probes or space shots.

Space probes are rockets sent up from Earth. They carry cameras to photograph objects in space. They also carry instruments to take scientific measurements. These space probes then radio the photos and measurements back to Earth.

American scientists launched the first space probe to Jupiter on March 3, 1972. They called the probe *Pioneer 10*. It had to travel 400 million miles from Earth to Jupiter. *Pioneer 10* took nearly two years to make the trip. It flew within 81,000 miles of the surface of Jupiter.

The *Galileo* is one of several American space probes that have taken photos of Jupiter.

The space probe studied the gases surrounding the planet. It measured its magnetic field. And it sent its findings to astronomers on Earth by radio.

Over the following years, astronomers launched more probes of Jupiter — *Pioneer 11* in 1973, *Voyager 1* and *Voyager 2* in 1977, and *Galileo* in 1989. These space probes took photos and measured features of the planet. They radioed their results to Earth. The photos and figures gave astronomers much new information about this fascinating planet.

The space probe *Voyager 1* began its journey to Jupiter on September 5, 1977. It was launched atop a Titan rocket from the Kennedy Space Center in Florida.

The colorful clouds that surround Jupiter are constantly moving, propelled by storms and fierce winds.

Above Jupiter

Jupiter is not surrounded with air like Earth. The air around Earth makes plant and animal life possible. While no one is sure, most astronomers believe there is no life on Jupiter. If there is any form of life, it is probably some kind of tiny, microscopic being.

Instead of air, a thick layer of clouds covers Jupiter. The clouds wrap the entire planet like a blanket. Astronomers guess that these clouds may be 11,000 miles deep.

Jupiter's clouds are not fluffy and white like clouds on Earth. They are formed into brightly colored bands — orange, red, white, and tan. The bands are like belts that completely circle the planet.

Jupiter's colorful clouds are not calm, either. Wild storms continually rage in these clouds. The winds whip around at high speeds. Astronomers think that bright flashes of lightning shoot out in all directions. They also believe that thunder roars endlessly on the planet.

The clouds around Jupiter are very, very cold. The temperature is about 240° *below zero* Fahrenheit (F). Compare that to the clouds around Earth. Their average temperature is about 50° F *above zero*.

Jupiter's clouds are not bits of water or ice floating in the air like the clouds on Earth. The clouds around Jupiter consist of frozen bits of gases. The two most common gases are ammonia and methane.

On and Inside Jupiter

Unlike Earth, Jupiter does not have a solid rock surface. Jupiter is mostly made of gases. Stepping onto the surface of Jupiter would be like stepping onto a puff of air. Most of the gas in Jupiter is hydrogen. There is also some helium.

Jupiter's powerful gravity pulls on this gas. It pulls the gases toward the center of the planet. The gravity keeps the gas together. It also causes tremendous pressure inside Jupiter.

When hydrogen is under pressure it changes from a gas into a liquid. This is what happens on Jupiter. The very strong pressure beneath the surface of Jupiter turns the hydrogen gas into liquid hydrogen. Most of the inside of Jupiter is hydrogen in liquid form.

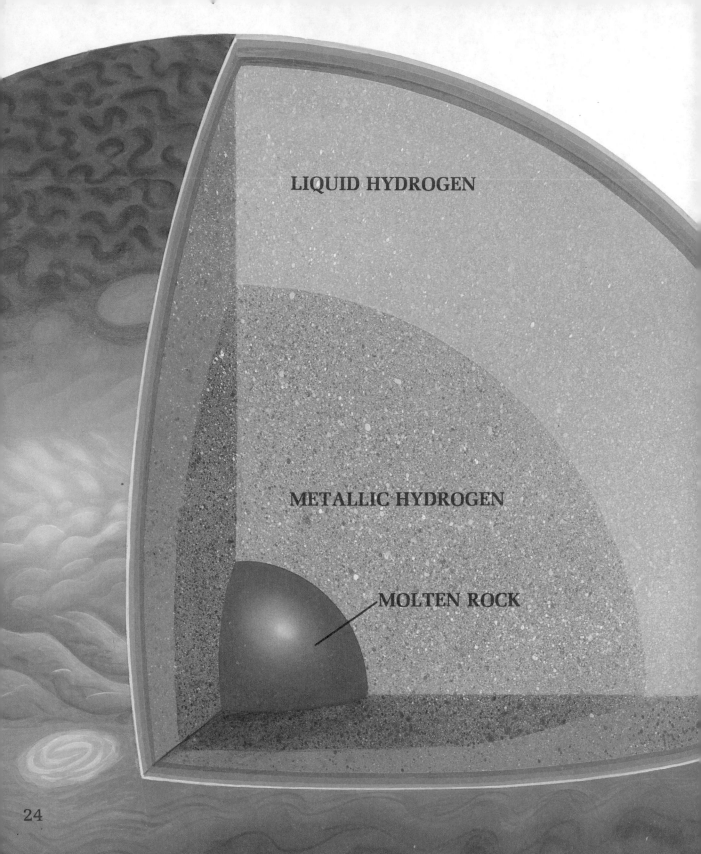

LIQUID HYDROGEN

METALLIC HYDROGEN

MOLTEN ROCK

The very center, or core, of Jupiter is solid. It is made of iron and rock like planet Earth. The core is also about the same size as Earth. But Jupiter is so huge that the core is just a small part of the planet.

Beneath the surface of Jupiter, the temperature rises very quickly. The core of Jupiter is far hotter than its surface. The core temperature may be as high as 43,000° F.

The gas, liquid, and solid parts of Jupiter are not very heavy. They weigh much less than the rock that makes up Earth. Imagine that you could fill a bottle with material from Jupiter. Suppose it weighs one pound. Now let's say you filled the same bottle with pieces of Earth rock. It would weigh about four times as much, or four pounds.

CLOUD TOPS
AMMONIA CRYSTALS
AMMONIUM
HYDROSULFIDE CRYSTALS
WATER ICE DROPLETS

Jupiter in Space

The planet Jupiter is always moving in orbit around the Sun. The time it takes to finish an orbit is a year on that planet. It takes Jupiter nearly 12 times as long as Earth to complete one orbit. Therefore, one year on Jupiter is the same as about 12 years on Earth.

If you lived on Jupiter, the long year would be very confusing. Each season of the year would last as long as three Earth years. And you'd have to wait 12 Earth years to have your first birthday party!

Like the Earth, Jupiter spins, or rotates as it orbits the Sun. But Jupiter rotates much, much faster.

The time it takes to make one rotation is a day on that planet. Jupiter rotates so fast that a day on Jupiter lasts less than 10 hours. A day on Earth, as you know, lasts 24 hours. The days are shorter on Jupiter than on any other planet in the Solar System.

The fast rotation affects Jupiter in odd ways. The great speed shapes Jupiter's clouds into bands. It also causes Jupiter to bulge in the middle. The planet looks like a ball that is flat on the top and bottom. This makes the distance around Jupiter's equator greater than the distance around the poles.

On Earth, you sleep about ten hours a night. If you lived on Jupiter, you'd only sleep about four hours a night because the days are much shorter. On Earth, you're in school six hours a day. On Jupiter, the school day would last only two and a half hours.

The Great Red Spot

The most striking sight on Jupiter is the Great Red Spot. The Great Red Spot covers hundreds of millions of square miles.

People first noticed the Great Red Spot more than 300 years ago. Since then, astronomers have watched changes occur.

The Great Red Spot changes size. It grows bigger and smaller. It also changes shape. Sometimes it is longer, sometimes rounder. And it changes colors, from bright red to pink to gray.

The Great Red Spot has even disappeared for short periods of time. But it always comes back in just about the same place.

Astonomers have long wondered what causes the Great Red Spot. Most now agree that it is a raging storm, like a powerful hurricane. Swirling clouds rush around at very high speeds. The color comes from different gases mixed in with the clouds.

The *Voyager 2* spacecraft made two important findings about the clouds within the Great Red Spot:

All the clouds within the Great Red Spot are not spinning around at the same speed. The lower ones are spinning faster than the upper ones. This means that something on Jupiter's surface is making the clouds spin. But so far the cause remains a mystery.

The lower clouds spin from west to east. But the higher clouds spin in the opposite direction, from east to west. The cause is also not known.

These photos, taken four months apart in 1979, show how Jupiter's clouds keep moving.

Photo A (left) was taken by *Voyager 1* in January. Look for the white oval below and to the left of the Great Red Spot. Now look at Photo B, which was taken in May by *Voyager 2*. The oval has moved eastward (to the right).

Three planets the size of Earth could fit inside Jupiter's Great Red Spot.

Jupiter's Moons

Most planets have moons in orbit around them. Jupiter has 16 moons. That is far more than any other planet. The Earth, as you know, has only one moon. Many astronomers think there may be even more than 16 moons that orbit around Jupiter. They just have not yet been seen.

The first person to discover moons around Jupiter was the Italian astronomer Galileo Galilei. In 1610, Galileo saw four moons with his early telescope. Through his telescope the moons looked like little specks of light.

We call the four moons seen by Galileo *Galilean moons*. The Galilean moons are the largest of Jupiter's 16 known moons. Each one is between 2,000 and 3,000 miles wide. They are about the same size as the Earth's moon, which is just over 2,000 miles across.

The closest Galilean moon is a little more than 250,000 miles from Jupiter. That is a bit greater than the distance from Earth to Earth's moon. The farthest Galilean moon is more than a million miles away from the planet.

The 12 other moons of Jupiter are all very small. They are no more than 100 miles across. Galileo couldn't see them with his primitive telescope.

Separate photographs of Jupiter and its four largest moons were put together for this picture. The moon at the bottom right looks larger than Jupiter because it is placed in the front. It is actually much smaller than the planet.

Four of these moons are closer to Jupiter than the Galilean moons. The eight other small moons are farther out. The farthest moon is nearly 15 million miles from Jupiter.

Most of the moons orbit around Jupiter the way our moon orbits around Earth. That is, the moons move from west to east. But the four farthest moons move in the opposite direction. They move from east to west.

No one is quite sure why these four moons of Jupiter move backwards. Some think that the four moons were asteroids that were captured by Jupiter's gravity. And now Jupiter's powerful pull keeps these moons in this odd orbit.

Io

The Galilean moon closest to Jupiter is called Io (EYE-oh). It is a little farther from Jupiter than our moon is from Earth. It is also a little bigger than our moon.

Voyager 1 and Voyager 2 made two important discoveries about Io. First, Io has active volcanoes. In fact, it is the only moon in the entire Solar System known to have volcanoes that are now erupting. The Voyager space probes located about a dozen volcanoes on Io. Astronomers believe there may be many more.

Some astronomers call Io "the pizza moon." Can you see why?

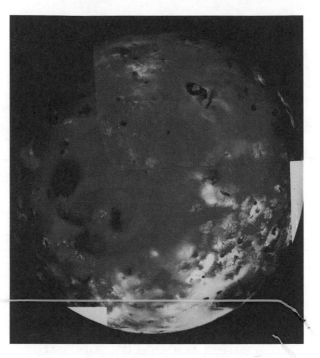

Io's volcanoes seem to be larger than the volcanoes on Earth. The volcanoes pour out hot, molten sulfur. The sulfur covers the entire surface of Io. It gives the moon its reddish-yellow color.

Also, Io has no craters. Craters are giant holes made when objects from space — comets, meteorites, or asteroids — slam into a larger body. The falling objects dig craters in the surface.

The planets Mars and Mercury, and our moon, are covered with craters. (Earth, too, has a number of craters. But wind, rain, and growing plants usually fill in the Earth's craters. This makes them hard to find.)

Astronomers wonder why Io is free of craters. They have two possible answers. Old craters may have been filled in by the liquid sulfur from Io's volcanoes. Or, Io may be a young moon and not many objects from space have crash-landed there yet.

Europa

We call the smallest of the Galilean moons Europa (you-ROH-pa). Europa is probably made of rock covered by a thick layer of ice, about 60 miles deep. Thin dark lines run in all directions across its icy surface.

Astronomers believe that the dark lines on Europa are cracks in the ice. The cracks are not empty. They are filled with dark rocky material from beneath the ice.

Early in 1995, scientists using the Hubble Space Telescope made an amazing new discovery about Europa. The moon is surrounded by a very thin atmosphere of oxygen. Four other moons in the Solar System have an atmosphere. But Europa is the only one with the same form of oxygen as that on Earth.

The presence of oxygen does not mean that there could ever be life on Europa nor does it mean that astronauts could ever walk on Europa without space suits. With temperatures of -230° F, it is far too cold for anything to live there.

Europa, the smallest of the four Galilean moons, is about the size of Earth's moon.

Ganymede is the largest moon in the entire Solar System. It's even bigger than the planet Mercury!

Ganymede

Ganymede (GAN-uh-meed) is the biggest Galilean moon.

Astronomers can see many craters on Ganymede. The craters were probably caused by giant hunks of rock or metal, called meteorites, smashing into the moon. Other parts of Ganymede show grooves. These came from breaks in the surface that filled with water. The water then froze and formed icy ridges.

The moon Callisto is more than one million miles from Jupiter.

Callisto

Farthest from Jupiter is the Galilean moon Callisto (ka-LISS-toe). Craters made by falling meteors dot the surface of Callisto. In fact, Callisto has more craters than any other body in the Solar System.

Some of Callisto's craters have rings around them much like a bull's eye. Astronomers believe the rings may have come from the impact of meteorites. The meteorites sent out waves when they struck. The waves are much like the rings that form when you throw a rock into water. Since Callisto is so cold, the rings may have frozen in place.

Ring Around Jupiter

The *Voyager 1* spacecraft discovered something else when it flew past Jupiter. A ring goes all around the planet.

Jupiter's ring starts about 34,000 miles from the top of its clouds. The ring is very thin. Astronomers guess that it is only 18 miles thick. But they think that it is about 4,000 miles wide.

The ring around Jupiter consists of tiny rocks and bits of dust circling the planet. Astronomers are not sure about the source of the dust. Some say the dust comes from the erupting volcanoes on the moon Io. Others believe a passing comet or meteoroid dropped the dust.

No one knew there was a ring around Jupiter until it showed up in photographs taken by *Voyager 1* in 1979. A thin line has been drawn on this photograph to show the position of the ring, which is too faint to see here.

The planet Saturn is farther from Earth than Jupiter, but its rings can be seen through telescopes on Earth.

About Comets

The amazing collision of 1994 occurred when a comet slammed into Jupiter. But what exactly is a comet?

Comets are objects in space. They usually move in orbit around the Sun like planets. But comets are much smaller than planets. Each comet consists of a small, solid nucleus, or head. Behind the head is a long tail of gas and dust.

An astronomer once described the head of a comet as a "dirty snowball." It is made up of ice, with chunks of metal and rock frozen inside. The chunks may be as small as grains of sand or as big as walnuts.

Two famous comets from the past are shown on these pages. They are the comet Kohoutek (1974) (above) and Halley's comet (left), which appears every 76 years.

To picture a comet, imagine a giant scoop of chocolate-chip ice cream in a cone. The scoop of ice cream is like the frozen head of a comet. The chocolate chips are like the rock and metal bits frozen inside. And the melted ice cream dripping down inside the cone is like the comet's tail.

Astronomers believe that the bits of rock and ice are leftovers from the birth of the Solar System. Most comets are found in the farthest reaches of the Solar System. But once in a while a comet wanders into the interior of the Solar System. Then the comet's path may collide with a planet.

Comets give astronomers much information about the origin of the universe. They tell scientists about conditions in outer space. And they provide clues to the future of our solar system, including our planet Earth.

Halley's comet was last seen in 1986.

Comet Hunters

The husband-wife team of Eugene and Carolyn Shoemaker, and David Levy, are three well-known comet hunters. Between them they have discovered nearly 50 comets.

The three astronomers decided to look for comets on the night of March 23, 1993. They worked as they had many times before. They used a telescope on a dark mountaintop in California. As usual, they attached a camera to the telescope to take photos of the sky.

At first, much of the sky was covered with clouds. So they waited and didn't take any pictures. Around midnight the clouds thinned out a bit. The team loaded film in the telescope. And they took some pictures through the thin clouds.

When the photos were ready, Carolyn Shoemaker started to study them. She saw all the dots she expected to see. But she also saw something unexpected. There was a streak on the photo.

A streak usually signals a comet flashing across the sky. But this one looked different. "It looks like a squashed comet!" she called out in great excitement.

Levy sent a computer message to the organization that keeps track of comet discoveries. He told the exact position of the comet they had seen. The officials could find no record of a comet at that spot.

Eugene and Carolyn Shoemaker and their partner, David Levy, were the first ones to see the comet that was about to crash into Jupiter.

Levy also asked another astronomer to look for the same streak. The astronomer agreed it was a comet that had not been seen before.

But the comet puzzled the experts. It looked different from other comets. It didn't have a single, solid head. This comet looked like a string of pearls. The astronomers guessed that the largest "pearl" was about a mile wide. The rest ranged down to a width of about a half-mile.

In time, they learned more about the comet. It had about 20 separate parts. The parts were spread out in a line over some 2 million miles of space.

The Shoemakers and Levy were the first astronomers to see the comet. So the comet was named Shoemaker-Levy. It was also the ninth comet the group had discovered together. The astronomers named it Shoemaker-Levy 9.

Comet on a Collision Course

Scientists now tried to learn more about Shoemaker-Levy 9. They figured out that the comet had been in orbit around the sun for many millions, or even billions, of years. It looked like other comets. Then, sometime in the early 1970s, this comet was captured by Jupiter's powerful gravity. It was pulled into orbit around Jupiter. In July 1992, the comet came within some 16,000 miles of Jupiter. The planet's strong gravity ripped it into many smaller pieces. The pieces made up the "string of pearls" seen in the photo.

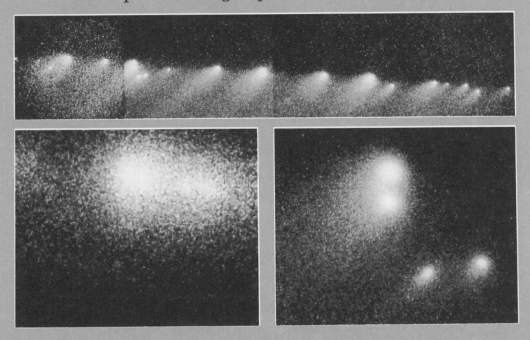

In the top picture you can see the comet's "string of pearls." Each bright spot is a fragment of the comet. The two shots at the bottom are enlargements of some of the fragments.

Astronomers carefully tracked the path of Comet Shoemaker-Levy 9. They wanted to know where it was headed. They were astounded at what they found. The comet was on a collision course with Jupiter! The first "pearl" would strike the planet on July 16, 1994. The others would follow over the next six days.

Everyone wanted to watch the event. But there was a problem. Shoemaker-Levy 9 would hit on the side of Jupiter facing away from Earth. Observers on Earth would not be able to see the impacts.

Nevertheless, astronomers trained their biggest telescopes on Jupiter. They knew that the planet rotates very fast. Perhaps the collision site would spin into view soon after the comet struck. Then the effects might still be seen. Or maybe there would be giant explosions when the comet landed. The flashes might be visible from Earth.

Luckily, three spacecraft were in position to view Jupiter from outer space. Closest was the *Galileo* space probe. The camera on board *Galileo* could take photos of the event. But there were problems.

Galileo was 121 million miles away. It had trouble "seeing" anything that was less than a few hundred miles across. Also, *Galileo*'s main antenna was broken. It would take months for *Galileo* to send back any information.

Astronauts make repairs to the Hubble Space Telescope, one of the spacecraft that sent back information about the comet collision.

Next closest was the Hubble Space Telescope. It was 480 million miles from Jupiter. The Space Telescope could not see the place where the comet chunks struck. But it could measure changes caused by the impacts.

Finally, *Voyager 2* was outside the Solar System, nearly 4 billion miles from Jupiter. Yet it could see the impact sites directly and measure changes. And it could radio its findings down to

Crashing Into Jupiter

Everything happened as expected. The first part of the comet, called Fragment A, crashed into Jupiter just after 4:00 P.M. (Eastern Daylight Time) on the afternoon of July 16. This fragment was about one mile wide. It was as though a mountain had smashed into the planet at a speed of over 130,000 miles an hour!

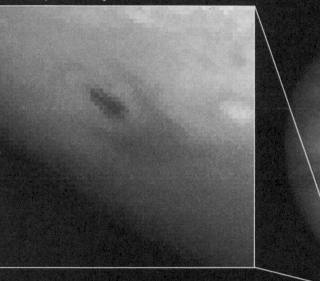

A Impact · July 16, 1994 · 19:00 UT

The Hubble Space Telescope was not in position to "see" the comet fragments crashing into Jupiter, but it was able to photograph the places that were hit. The dark area shows one of the impact sites.

The huge chunk did not tunnel very deeply into Jupiter's atmosphere. Probably it did not reach below 40 miles or so. But the crash caused a powerful explosion. The blast was as great as 50 hydrogen bombs exploding all at once! The explosion was on the far side of Jupiter. Yet astronomers could see a flash of light from Earth.

About 12 minutes later, the impact site rotated into view from Earth. Astronomers on Earth saw a huge fireball above the surface of Jupiter. This fiery plume looked like a geyser that had been shot up out of a giant cannon. The plume was made up of white-hot gas and dust. It stretched over 1,000 miles high and 4,000 miles across.

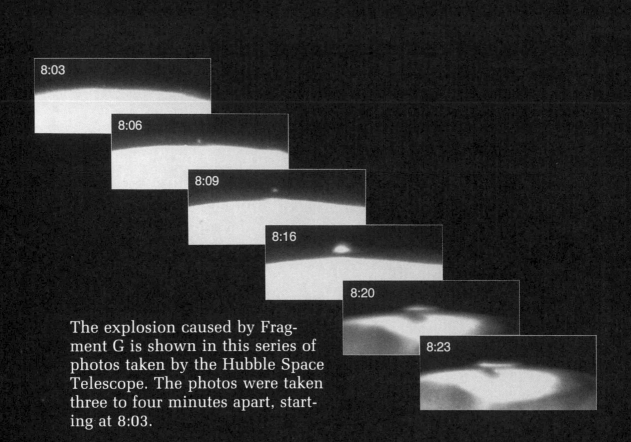

8:03

8:06

8:09

8:16

8:20

8:23

The explosion caused by Fragment G is shown in this series of photos taken by the Hubble Space Telescope. The photos were taken three to four minutes apart, starting at 8:03.

After about 15 or 20 minutes, the fireball faded away. But it left a dark splotch 7,000 miles wide on Jupiter's atmosphere.

Over the next six days, 20 more chunks of Shoemaker-Levy 9 smashed into Jupiter. The collisions sent up huge columns of gas and dust. In a little while the plumes were gone. But they left giant black spots on the planet. Astronomers could see the spots again and again as Jupiter spun around.

H B N Q1 Q2 R D/G   L

In this picture, each letter stands for a fragment of the Shoemaker-Levy 9 comet that struck Jupiter. The arrows point to the areas that were damaged.

The biggest chunk of the comet was Fragment G. Fragment G was about two miles across. The crash of Fragment G made a fireball 1,600 miles high and 5,000 miles wide. The fireball was bigger than the entire planet Earth!

The flash of light was blinding. The light was brighter than anything that had been seen before. Instruments on Earth telescopes could not even measure its brightness.

The Results

The spectacular sky show was over in a few days. But it gave astronomers facts and figures to study for years. Scientists in laboratories all over the world are still gathering valuable information from their observations.

At present, experts are hard at work analyzing the data. They hope to use some of the information to answer such questions as:

What gases are in Jupiter's atmosphere?

What material is inside Jupiter?

What rocks and metal were in the comet?

How deep into the clouds did the comet pieces fall?

What makes the colors of Jupiter's clouds?

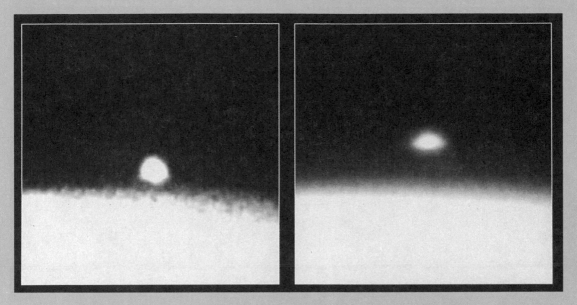

Fragment G was the biggest chunk of Shoemaker-Levy 9. Above you can see the fireball coming out from Jupiter's surface. The two pictures were taken by the Hubble Space Telescope, which was 480 million miles away.

Below is a close-up of the impact site after the fireball.

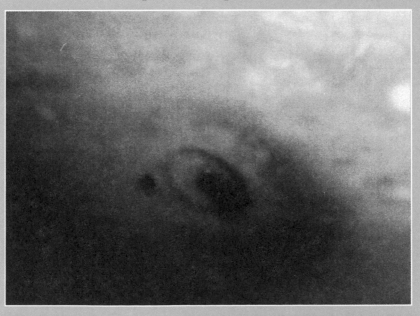

The comet-finders celebrate "Nature's home run."

David Levy said this about the heavenly show: "Nature has hit us a home run. . . ." The "home run" will add to what we know about outer space. It may tell us more about the origin of the Solar System. It may even suggest what will happen if a comet strikes Earth.

Collisions With Earth

Astronomers know that comets sometimes crash into planets — including Earth. On average, a really large comet lands on Earth about once every 100 million years. Less serious collisions may occur every few hundred years.

The biggest comet blast on Earth took place probably about 65 million years ago. Remains of a huge crater have been found in Mexico's Yucatán Peninsula. Scientists believe that the explosion sent up an immense dark cloud of dust. Winds carried the dust all over the world. The dust blocked the sunlight so that plants could not grow. Without plants, the dinosaurs had nothing to eat. In this way, many think, that particular comet collision led to the extinction of the dinosaurs.

When a comet or other object from space hits the earth, it makes a crater in the ground. This crater in Arizona was caused by a meteorite.

An object from outer space — perhaps a comet — landed in Siberia in 1908. The blast flattened trees up to 100 miles away. The powerful explosion caused great damage. But it fell far from inhabited areas. No lives were lost.

What about future comet collisions with Earth? Scientists think that they may be able to prevent them. Suppose a comet is heading toward Earth.

Standing inside a crater is an amazing experience.

Astronomers will probably know about it months or years in advance. They may rocket a bomb up into space. They may explode it near the comet. Just a little nudge can send the comet millions of miles away from Earth.

Throughout the ages, astronomers have learned a lot about certain comets, such as Halley's comet, shown here in a photograph taken in 1910.

Collisions with comets can cause great harm. But they can also be a good force of change. In fact, life on Earth may have been started by a comet entering the Earth's atmosphere from outer space. The comet that struck Earth might have been carrying some microscopic living beings. Perhaps the organisms landed on Earth and began to grow. And over billions of years they developed into the many different kinds of plants and animals on Earth today.

Collisions between comets and planets are incredible events! Just like the crash of Shoemaker-Levy 9 on Jupiter, comet collisions have amazing possibilities.

Scientists hope that the Hubble Space Telescope will help them learn even more about Jupiter and the planets beyond.